ENTERPRISE RESOURCE PLANNING (ERP)
THE GREAT GAMBLE

ENTERPRISE RESOURCE PLANNING (ERP)
THE GREAT GAMBLE

An Executive's Guide to Understanding an ERP Project

Ray Atkinson

Copyright © 2013 by Ray Atkinson.

Library of Congress Control Number: 2013909238
ISBN: Hardcover   978-1-4836-4443-1
                Softcover   978-1-4836-4442-4
                Ebook   978-1-4836-4444-8

All rights reserved. No part of this book may be reproduced or transmitted in any form or by any means, electronic or mechanical, including photocopying, recording, or by any information storage and retrieval system, without permission in writing from the copyright owner.

Rev. date: 07/18/2013

To order additional copies of this book, contact:
Xlibris LLC
1-800-455-039
www.Xlibris.com.au
Orders@Xlibris.com.au
503778

CONTENTS

A Brief History of the Evolution of ERP 11

ERP Disasters 16

The Blame Game 19

The ERP Industry 25

The 26 Steps to successfully understanding an ERP project 32

    1. ERP Acquisition Rationale 35

    2. In-House Expertise in ERP 36

    3. ERP Risk Assessment 37

    4. ERP Cost Justification and Budget 38

    5. Request for Proposal 40

    6. Software Selection Criteria 44

    7. The ERP Contract 46

    8. Project Plan 48

    9. Go-Live Date(s) 49

    10. ERP Education 50

    11. ERP Training 51

    12. Implementation Responsibility 53

    13. Project Management 55

14. Executive Involvement 56

15. Software House Expertise 57

16. Software Implementation Team Expertise 62

17. Process Change 63

18. Data Clean-up 65

19. Data Conversion 66

20. Issues Identification 67

21. Scope Change 68

22. Software Changes 69

23. Management Action on Issues 71

24. Go-Live Readiness Reviews 72

25. Live Running Cutover 75

26. Post-ERP Cutover 76

Those who do not learn from history are doomed to repeat it!

(George Santayana)

Foreword

ENTERPRISE RESOURCE PLANNING (ERP) has been around since the late 1960s. Originally, it was called MRP, then MRPII, and in the 1990s was rebadged as ERP. The concept of ERP is the utilisation of the power of the computer to integrate all the functions of a business in order to provide real-time information for planning and decision-making.

The transition from manual systems to computerised systems has not been an easy task for the majority of companies. Surveys regularly show high rates of failure or underperforming systems with costs exceeding budgets, time delays, and major disruption to business activities.

For too long, some suppliers in the ERP industry have hidden behind a veil of deceit, misinformation, and dubious consulting practices that are more attuned to filling the coffers of the ERP vendors and their consulting teams/partners than providing successful outcomes for the ERP-buying customers.

With the huge sums of money involved in ERP projects and the high rate of failures and underperforming systems, the rise in the number of legal actions against ERP vendors and their partners is a reflection of the unwillingness of organisations to pay large sums of money to software vendors and their partners who make a mess of the implementations and then walk away unscathed, frequently blaming the customer.

It is inconceivable that 35 years of failure rates of around 70% (well researched) that we, as experienced practitioners, cannot develop a process for implementing ERP that provides cost-effective outcomes most of the time. It may be that with so much money to be taken from customers, there is more to be gained by the ERP industry in leaving the status quo and walking away post-failure and accepting no responsibility for the outcome.

Our corporations and industries cannot afford these disasters. ERP projects drain resources, distract management, and become liabilities that have little returns for the buying clients.

This book is *not* a technical manual explaining all the nuts-and-bolts details of ERP that must be mastered to successfully implement the technology but is a guide to senior executives, managers, project managers, and project teams to understand the different aspects of an ERP project.

An ERP project is far broader than the software technology and it is these other issues that can be the difference between success and failure.

This book is based on 35 years of experience of the author, who has worked in organisations all over the world in various capacities and has project-managed ERP projects with varying degrees of success and failure and has analysed many ERP projects from a recovery, mediation and litigation perspective to determine the underlying reasons for ERP failure.

The book is written in layman's terms and seeks to provide senior management, middle management, project management, and their project teams with an understanding of the issues that need to be addressed and managed in order to achieve a successful outcome from an ERP project.

CHAPTER 1

A Brief History of the Evolution of ERP

ENTERPRISE RESOURCE PLANNING (ERP) came about with the introduction of computer systems that could be utilised by employees to perform their functions a lot faster and more accurately across a broad range of areas within the organisation.

In the 1960s–1970s, the technology was called materials requirements planning (MRP). This was primarily developed for the manufacturing industry and where multiple parts and raw materials were required to manufacture products. MRP was far superior to the min/max calculation, which had been the only effective means of materials planning for mass production before computers. The MRP calculation involved a list of materials that were required to build the product (bill of material/BOM), a quantity of products to be manufactured, and a link to the inventory records to determine what materials were in stock and what needed to be purchased in order to build the product. There were limitations to this as the calculation provided a list of materials but not exactly when the materials were required.

In the 1970s, users recognised that the system lacked the ability to indicate when exactly the parts were required and they incorporated into the software a Master production Schedule (MPS) and the sequencing of steps to build the products though the production works centres called Routings which provided a specific time the materials were required in production. These additional developments enabled the purchasing departments to plan and schedule the exact timing parts were due to support production whilst keeping inventory in synchronisation with actual requirements. In addition a rough cut capacity planning facility was added to enable a quick sizing of capacity availability based on families of product as a means of providing some validity to the master production schedule and providing the capability of the system to predict when product would be available to promise (ATP) to customers.

Over time, other enhancements were the linking of sales order entry, detailed finite capacity planning, product costing, finance, maintenance, despatch, and invoicing into an integrated set of modules that provided real-time management decision support information. The name given to the enhanced system was upgraded from MRP to Manufacturing Resource Planning (MRPII)

A key element of the development of the technology was the recognition that the technology had to be used in a certain sequence to be effective. This became known as the 'MRPII philosophy', which required an education program for the organisation to understand how to use the technology and embrace the changes in the way the business needed to be operated for the technology to be a success.

At a theoretical level, the technology made a lot of sense, but implementation of the technology was problematic with only 30% of companies reporting successful outcomes. There were various reasons for this! Primarily was the tendency to believe that MRPII was a computer system and the role of education was a step that could be bypassed in the belief that the employees were smart enough to pick it up as they went along. There were many other issues that undermined MRPII projects including the tendency to look upon the technology as a computer system that could be relegated to the computer people, the failure to understand the philosophy, the lack of accuracy of the available data to support the system and the problems with selecting systems based more on accounting requirements than the other operational needs of the business.

In the 1990s, MRPII was rebadged as enterprise resource planning (ERP) and was touted as the successor to MRPII. The reality was MRPII packages were rebadged overnight into ERP with no discernible difference from the original MRPII. It became more of a marketing exercise to stir up customer interest and get away from the perception of the MRPII failures that were common knowledge and to make the technology and its integration characteristics more appealing to the non-manufacturing sector, thus the expansion of the technology into other areas such as human resources (HR), projects, maintenance, customer relationship management (CRM), etc.

With the change of name and the expansion of the technology into other areas of business, an industry has grown up around ERP of software providers, software and technology add-ons, consulting partners specialising in implementation, and a myriad of ERP experts. The global ERP industry is worth around $50 billion! The MRPII failures of the 1980s and 1990s were disruptive and frustrating, but basically the costs were contained to software acquisition, some education, training and configuration, and some ad hoc consulting. Rarely was it seen as potentially sending organisations bankrupt or significantly impacting financial results.

With the introduction of the ERP tag, software companies saw a business opportunity in offering project and implementation services to project-manage the technology into organisations, for a fee (a very large fee), and pushed this as part of their sales pitch, basically saying, 'We'll do the work for you'.

Many software companies created separate implementation divisions within their companies or partnered with some of the large consulting firms. These project and implementation services sometimes carry huge price tags, unjustified for the service provided and the results achieved. In addition, internal projects, within organisations, were established to support the implementation effort, and in many cases, these were remote from the people who would end up using the technology. These teams were typically orchestrated by the software companies and their partners, and the project plans were couched in technical jargon far beyond the client company's comprehension, in order to justify the costs involved. The implementation services were packaged in such a way that it led CEOs and senior executives to believe that ERP can be bought by paying a large amount of money and be left to someone else to implement it. We will prove this to be a totally wrong approach.

The cost of these add-on services can run into hundreds of thousands and millions of dollars, and companies were and are willing to pay this in the belief that they will achieve a good return on the investment through the gains the technology offers.

Constant surveys show that successful ERP projects run around 30–45% of all implementations. In addition, most implementing companies report (!) an average of 25% overruns on the budgeted costs. We consider this figure to be wildly optimistic. This is despite paying large sums of money for implementation services. Success is difficult to gauge as in the eyes of some, just turning it on to stop the cash bleed is success.

The figures on return of investment are also difficult to assess, but surveys would suggest that companies reporting a successful implementation are only achieving around 50% of the expected benefits. Many companies are unable to measure any discernible benefits, but this may be due to the non-existence of a clear definition of what the ERP goals are in the first place.

Companies are reluctant to disclose too much information about their ERP projects as they either feel embarrassed about their failures or do not wish to reveal problems as it may reflect badly on management and affect their brand image.

The true cost to our industries and organisations across all sectors can only be surmised as we typically only hear about the high profile disasters and receive information on surveys that consistently draw negative conclusions on the ability of organisations to successfully implement the ERP technology.

Having said all the above, there is no doubt that successful ERP systems can and do provide huge benefits to an organisation's ability to plan and control their operations by being able to access real-time information to support decision-making. This dream will only be realised for the very few who are able to master the process of implementing an ERP project by understanding the scope and impact of these types of projects and take a much broader view than ERP being simply a technology project. Decisions made by senior executives at the very beginning of ERP projects have a profound impact on success or failure of the project, but the impact of these decisions on the success or failure of projects is poorly, if at all, understood.

Chapter 2

ERP Disasters

THE HISTORY OF MRP and ERP systems is littered with high-profile public disasters. The failures we read about are only the tip of the iceberg! The reality is that most ERP systems implemented do not live up to the original expectations. 'Original expectations' is what was or should have been clearly spelled out in the original cost justification for the project! For companies attempting ERP projects complaints of massive cost overruns, project time schedule extensions, inadequate software functionality, negative impact on employee morale, little benefit for the investment, lost business opportunities, negative impacts on financial results, damage to brand image and, in some cases, bankruptcy have been cited as outcomes they have experienced.

Some examples of bad ERP outcomes:

Hershey reported a 19% drop in third-quarter net earnings and attributed the fall to a botched ERP implementation.

Whirlpool blamed shipment delays in part to its SAP implementation which went live 2 months prior. Apparently, orders for quantities smaller than one truckload had faced snags in the areas of order processing, Tracking, and invoicing. SAP reportedly said Whirlpool went live with the software before adequately testing it and preparing their supply chain.

Invacare lost $30 million as a result of a bungled Oracle ERP implementation.

W. L. Gore & Associates filed a lawsuit against consulting firm Deloitte & Touche and PeopleSoft over an allegedly botched attempt to install PeopleSoft's HR module.

Nike's $400 million upgrade to their ERP system cost them $100 million in lost sales and 20% drop in their stock value and generated a collection of lawsuits.

Hewlett Packard's centralisation of its disparate North American ERP systems on to one SAP system cost them $160 million in lost sales and order backlog.

Waste Management ended up getting embroiled in an acrimonious $100 million legal battle with SAP over an 18-month installation of its ERP software. The initial deal began in 2005, but the legal battles commenced in March 2008, when Waste Management filed suit and claimed SAP executives participated in a fraudulent sales scheme that ended up in a massive failure.

Select Comfort, bedding maker, was forced by shareholder pressure to end their $20 million plus ERP project on the basis of 'extremely poor judgment by management' (charged one shareholder's SEC filing).

Ansell, Australian condom and medical protection giant, botched an implementation of Oracle's Fusion ERP platform which caused US$13–US$15 million worth of lost sales. It is hard to quantify the impact of Fusion, but lost sales in (the first financial half) were estimated at US$13–US$15 million, and it generated an excess working capital requirement of US$25–US$30 million.

These few examples of major ERP failures are from significant organisations that have the resources to tackle large ERP projects. There are many examples of smaller organisations who do not hit the headlines being severely and financially damaged by ERP projects that have ended up as failures.

For many organisations, ERP projects start out with great optimism as to the benefits that will derive from embarking on the project. For the most part, these projects drain resources, are poorly planned, woefully underfunded, and end up exhausting the organisation's resources, morale, and confidence in the technology and even management.

Typically, organisations are simply happy to turn on half-completed projects to stop the cash drain and get something from the technology. An ERP information system turned on, where critical steps have been omitted or half-completed, beggars the imagination as the system is supposed to provide information for decision-making. The result is usually internal chaos, lost business opportunities, and badly burnt executives that entrusted the project to others. Various outcomes can be achieved from this but fall far short of the technology capability, potentially leaving the company in a disadvantageous competitive position.

Chapter 3

The Blame Game

THERE ARE MANY players in any ERP project. Senior management, who approve the project on a variety of criteria, middle management, who are typically tasked to implement the technology or upgrades, various internal or external experts, the software seller and their implementation partners, and the ultimate end-users.

The majority of, if not all, ERP implementations end up underperforming against the original expectations! In many cases, expectations are that the technology will fix everything! The extent of the underperformance will determine the level of the blame game. In the past, it has been the practice of software houses and their partners to point to the many failures of the implementing companies as the main cause of blame for a failed ERP project, and there are plenty of examples to demonstrate this. Software companies rarely focus on their own failures as this could bring them into the legal liability arena and affect future sales.

The definition of failure depends on which part of the project you sit on. From the ERP supplier's and their implementation partner's perspective, success is, if you switch it on, the software runs. From the implementing customer's perspective, they are looking to achieve the outcomes for which they cost-justified. If there isn't a cost-based justification, some organisations even live with a pious hope that all their problems will miraculously be fixed. We know it doesn't happen . . . If you can't measure it, you can't manage it!

The cost of ERP failure has risen exponentially over the last 25 years, and it is not surprising to see the level of litigation increasing from disgruntled ERP purchasers. Claims of 'proven paths', 'best practice', and simplistic implementation methodologies that fail litter the ERP landscape as each software company seeks to gain some form of advantage over its rivals.

Typical issues from an ERP buyer's perspective:

- *Project overrunning budget and cost blowouts.*
- *Project is running late and poor ongoing progress.*
- *Taking short cuts to bring project back on time.*
- *Resource issues and lack of expertise.*
- *Internal confusion and conflicts.*
- *Management frustration.*
- *Little perceived benefit going forward.*
- *Wishful thinking by issuing instructions that ignore reality.*

The following issues are usually condensed into a list of complaints common to the majority of ERP projects.

Typical complaints from a company's perspective:

- *The ERP software vendor and their partners misrepresented the extent of their expertise and experience.*
- *The vendor assigned inexperienced people to the project.*
- *The claims of proven path and best practice implementation methodologies were misleading sales hype.*
- *The training was carried out too early and by inexperienced trainers who couldn't answer fundamental questions regarding software functionality.*

- *The software house concealed or misled the company on the limitations of the ERP software functionality.*
- *The software house and their implementation partners are more interested in gouging the company for additional dollars at every opportunity instead of implementing the system within the agreed timescale and budget.*
- *Language comprehension difficulties on the part of the trainers.*

The ERP software vendors and their implementation partners typically respond with a list of counterclaims that allege failure by the implementing company to address key issues such as the following:

- *Lack of top management commitment and involvement*
- *Poor initial requirements definition and ERP knowledge*
- *Budgeted cost underestimated for data clean-up, training, education, additional hardware, etc., and business disruption*
- *Lack of project implementation resources*
- *Scope creep*
- *Corporate behavioural and poor change management processes*
- *Unrealistic project implementation schedule*
- *Unrealistic software expectation versus reality*
- *Unrealistic expectation of benefits and ROI*
- *Inadequate training and education of users due to company's normal business pressures*
- *Unrealistic 'go-live' date(s)*

All of these issues and complaints have a degree of validity depending on the perspective of whether you are the software vendor, the implementing partner, or the implementing client. The reality is that all parties always share the responsibility for the successes and failures of the system. The term used to describe this is called *the Devil's Triangle*.

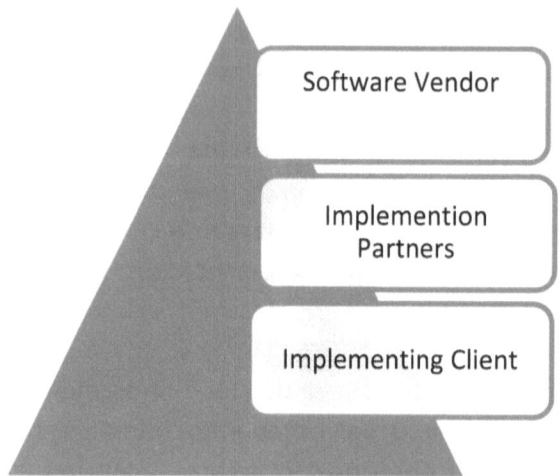

The involvement of the three parties in the ERP project lends itself to confusion and apportioning responsibility and blame for ERP cost and time overruns and poor outcomes. The finger can be pointed at all of the three parties involved, but ultimately, it is the client that ends up footing the bill and consequences for the organisation.

When ERP projects run into trouble or have a poor outcome and the blame game starts, it is important to remember that the implementing client is not an expert in implementing ERP systems as they purchase or upgrade systems perhaps once or twice over 20 years. On the other hand, the software house and their implementation partners have gone through this process numerous times, and it is reasonable to expect that they have the experience and skill in the implementation of the technology.

Remember, the vendor selling implementation services is holding itself out to be an expert in this area!

It seems sometimes that the ERP vendor and their implementing teams/partners have a greater expertise in apportioning blame to the implementing client than they have in identifying in-project problems and resolving them before they become a crisis.

With surveys indicating between 55% and 70% of ERP projects failing to produce expected results and most ERP projects suffering massive budget overruns, it is a reasonable assumption that software house and their implementation partners would have more expertise in project failure than success stories!

There is no other industry where buyers would pay so much money, experience such bad results, and would simply walk away without seeking recompense and compensation. The nature of the computing technology and the confusion that can be generated, obscuring clear lines of responsibility, create a perfect environment for unscrupulous ERP software suppliers and their implementation partners to walk away from failure, leaving the client with whatever mess they create.

Ethical vendors will not do this, but the buyer must learn how to talk to the supplier using the right language! Keeping vendors and their implementation partners honest by way of specific contracts, penalties, and milestones tied to payments can very quickly identify the ethicality from the charlatans.

The rise in litigation against ERP software vendors and their implementation partners indicates a change in attitude by the corporate world to holding those responsible for providing systems and alleged expertise, at great cost, to account for the misleading and deceptive conduct that plagues the IT industry generally.

The large sums of money involved, the promises of proven path and best practice implementation strategies from ERP vendors and their implementation partners, and the potential impact on the implementing client organisations of ERP failure ensure a growing tendency for injured ERP buyers to seek legal remedies for damage caused by misleading, deceitful conduct and fraudulent behaviour.

The days of 'taking candy from a baby' are coming to an end!

Organisations implementing ERP need to recognise their own shortcomings when looking at reasons for the failure of their ERP project. Over-reliance on the ERP vendor's claims of expertise, system capability, timescales, and the lack of senior executive scrutiny of the project plan and critical support when decisions need to be made outside the internal project managers mandate all contribute to the end result.

The blame game is the inevitable result of projects that have ill-defined areas of responsibility and are far more complex than just implementation of a computer system. It is hard to see much change in ERP outcomes if we continue to approach the issue as a technology issue rather than a business change process that has an ERP systems implementation as part of that process.

Chapter 4

The ERP Industry

THE GLOBAL ERP industry is worth in the region of 50 billion dollars! This is money taken from organisations that have a need for the ERP technology, and often outcomes are, at best, poor and can be damaging and in some cases can result in bankruptcy.

The ERP industry is dominated by ERP software vendors and their implementation partners, sometimes an arm of the same organisation, and their overriding objectives is to maximise the amount of revenue they can extract from every customer. This objective, of course, is not revealed to potential customers. Terms such as partnering, shared outcomes, knowledge transfer, etc. are all designed to make the potential customer believe that the ERP software supplier and its implementation partners have the company's interests at heart. Consistent survey results on ERP outcomes show a picture of an industry that is big on promises and very poor on results.

Despite the high level of failures of ERP projects, the software industry and their partners have shown no willingness to change a model that generates so much income for them . . . Why would they!

Our experience of multiple ERP projects consistently shows that ERP projects are not just about software but are about a massive change to the way businesses are run using computers to focus on changes to processes and decision-making.

The non-software components of ERP projects, which are poorly understood by either the buying company or the ERP software vendors and their partners, revolve around management decision-making and internal change that is beyond the authority and responsibility normally given to the ERP project and must be managed and driven, in detail, by senior executives.

The tolerance of organisations to ERP project failures is extraordinary given the amount of money invested in the technology. The involvement of the three parties to the Devil's Triangle generates sufficient doubt and confusion over liability that most organisations simply take the financial and ERP failure hit and muddle on with trying to get some return from the systems as installed.

In some cases, companies bleed for years attempting to implement ERP systems!

The ERP industry, generally, have gotten away with the behaviour that makes the mafia look honest. Products that don't do what the salesmen promise, expertise they claim to have but don't, misrepresentation, and creating chaos in industry with failed implementations . . . all these are commonplace.

There is no doubt the IT industry recognised a great business opportunity with the coming of integrated systems called ERP and have exploited it to the hilt. The additional implementation services now offered, at great cost, is a real revenue raiser for ERP vendors and their implementation partners, and given the inevitable delays to projects, they milk every last dollar out of the project, leaving the client with the legacy of whatever disaster they leave behind. In some cases, that can be financial ruin and bankruptcy!

The high cost of IT consulting services for ERP systems is in itself bad enough, but despite the huge-budgeted costs with overruns of an average 25% across all implementations, the success rate is abysmal. Around 55–70% of all ERP implementations do not provide the returns expected, and most implementations are seen as poor value for the money invested.

With more and more disputes arising out of the ERP industry and the buying companies prepared to seek legal redress against the software suppliers and their implementation consultants, one would think there would be a change in behaviour. Wrong! The rorts continue with the IT industry believing they can obscure their behaviour and blame the client sufficiently to avoid being held accountable.

The cost of the implementation of these ERP systems has escalated dramatically over the last 25 years to the point companies have been financially wounded or even sent bankrupt through botched implementations.

The failure rate for MRPII systems of 30 years ago was in the region of 70%. Failure is defined across a broad range of criteria such as delay in live running, cost overruns, projects abandoned, and reduced scope due to software and operational issues. Companies then and now reporting successful implementation only report 50% of benefits expected from the implementation of the technology. The difference is that the cost of failure has risen exponentially with the switch from internal client project ownership to the paying of money to software vendors and their partners.

The most common complaints from ERP customers against the ERP vendors and their implementation partners:

* *Selling software features and solutions that are far in excess of the requirements of the client.*

This is couched in terms of future proofing yourself against price increases and getting a bargain. In a lot of cases, the features and functionality will never be used, and the money is simply wasted.

* *Under-quoting for the software by not providing sufficient software functionality and then charging for the additional functionality during the implementation stage and blaming the client for not understanding the scope of the offer.*

This is a great revenue spinner as the client cannot operate without the additional functionality and has no choice but to pay up. In many cases, the client company contributes to this problem by putting forward a software budget without understanding that this may restrict functionality. The reality is that this only leads to scope creep later on when the realisation hits that the functionality is essential, leading to additional costs and delays.

* ***Offer experienced consultants***, *at a high dollar rate, as part of the support package and then provide inexperienced people at the same rate who may have some knowledge of software but have little or no clue on how to apply the technology, in a business sense. The client companies often become a training ground for new consultants at their cost.*

The company over-relies on the consultants and discovers the lack of expertise down the track when the project has gone horribly wrong only to find out that they, the client, are blamed for missed deadlines and that they can't rely on the details in the contract to claim recompense. Most consultants are very good in hindsight at telling the company what they didn't do and how they failed but are very poor in recognising and, more importantly, escalating the issues to management before they become a crisis.

* ***Project delay due*** to the client not having carried out specific tasks, causing additional consulting costs, as they wait for the work to be completed. The work was clearly outlined in the project plan, but the client didn't complete it in the time allotted.

The project plan couched in industry-invented jargon, put together by the consultants, is often unrealistic and based upon a go-live date, specified by the client, can be extremely optimistic and impossible to achieve. This is known to the consultants, who extract more revenue from the delay which, of course, is not their fault – even though they know, at the beginning, that the plan is unrealistic!

* ***The great modification rip-off*** is another money spinner whereby the software house carries out the modification at great cost and then retains the right to on-sell to other companies even though the client has paid for the work and should retain the intellectual property. This is also cited as a reason for project time delays. Software houses typically exempt modifications from ongoing maintenance upgrades as they have to modify the new upgrades to suit the modifications requested previously during implementation. Modifications are a great source of revenue for software companies. The lack of a model and the manner of demonstrations can lead the buyer to believe he has purchased the functionality only to find that to get the necessary functionality, modifications will have to be carried out, or additional software modules purchased, costing money and potential delays.

* ***The hostage situation,*** whereby when a company looks to alter their implementation strategy to reduce costs, as it is not working, the software house or their consultants create a fear environment by indicating that the new strategy will lead to failure and that the software house or

> *their consultants will not accept any responsibility for the consequences.*

The software house and their consulting partners do not take responsibility for project outcomes and certainly will not give up the revenue stream easily.

The ERP industry has evolved over the last few years where the over-reliance on software houses and their implementation partners has resulted in great swathes of money being passed over for, at best, dubious consulting services. With so much money being sunk into an ERP project, suggestions of failure are enough to put a scare into the toughest of CEOs. Ask the software house and their partners for a guarantee and watch them duck for cover!

With so many disasters in ERP projects in the community and the results from different surveys, it would be a reasonable generalisation to conclude that many software houses and their partners engage in activities that are questionable. The big boys of SAP, Oracle, Epicor, etc., seem to feature well in articles of organisations being sued, but they represent a fraction of the disgruntled ERP clients who have experienced high cost failures.

Whilst ever there is so much money involved in the ERP industry and with so little accountability from the ERP vendors and their partners, we will continue to experience the outcomes we see today, including the move towards litigation.

Undoubtedly, there are reputable ERP vendors and their implementation partners, but they are swamped by non-technology issues outside their scope and mandate.

Organisations seem to get better value from small independent or groups (1–5) of consultants that are not tied up with software providers and have no conflict of interest in protecting software vendors or trying to expand the scope of supply by on-selling additional software or services.

Separating the ERP systems sales, configuration, data conversion, and consulting functions provided by the software house and their partners from the ERP project management consulting function would seem to remove the basis for many of the questionable behaviour due to conflict of interests that take place today, and the sooner ERP buyers take control of their own projects and outcomes, through education, ownership, better contracts, and understanding of the technology, the sooner the ERP industry will clean its act up. This can only be driven by the client companies themselves as there is too much revenue at stake for the software vendors and their partners to change their behaviour.

Chapter 5

The 26 Steps to successfully understanding an ERP project

IN TRYING TO understand why so many ERP projects end up as underperforming and in some cases disasters that have sent organisations bankrupt, the author has analysed a number of projects over many years to find a common thread that leads to poor outcomes. There are many contributing factors to success or failure that cross a wide range of issues of which some are either ignored altogether as part of the project or are simply not understood or seen as relevant.

In an effort to understand the many issues that impact an ERP project, the following list of 26 key areas have been identified as critical to an ERP project but are poorly or not managed at all. Software houses tend to be experts in their own software and some peripheral issues, but they do not cover all the issues that impact success or failure of an ERP project.

These issues are listed here from 1 to 26! The list follows a logical sequence, and some can be switched about, but resolution of some issues needs to be resolved before others can be dealt with. For example, constructing a model after the software has been purchased simply leaves the company vulnerable to software that does not have the functionality to suit the operating processes of the company and may require massive customisation or additional software module purchase in order to meet the company's needs. These 26 steps take a global view of the company and what ERP is and what impacts success or failure.

Outlined in the following are the different essential steps involved in an ERP project and typically who manages the issues. This could be the buying company, the ERP vendor/team, or a joint responsibility between the company and the ERP vendor.

The reality here is that organisations do not give authority or mandate to any individual or group to manage all the issues identified as ERP is commonly seen, by senior executives and management as a technology computer project. The other areas are partially managed or not at all. ERP problems can be tracked back to the failure to resolve critical steps in the project that do not manifest themselves until the project runs into trouble. Following on is a brief explanation of the key points in each of the 26 steps.

Area	Responsibility
1. ERP acquisition rationale	Company
2. In-house expertise in ERP	Company
3. ERP risk assessment	Company
4. ERP cost justification and budget	Company
5. Request for proposal (RFP)	Company
6. Software selection criteria	Company
7. The ERP contract	Company/vendor
8. Project plan	Company/vendor
9. Go-live date(s)	Company
10. ERP education	Company
11. ERP training	Company/vendor
12. Implementation responsibility	Company/vendor
13. Project management	Company/vendor
14. Executive involvement	Company
15. Software house expertise	Vendor
16. ERP software imp team expertise	Vendor

17. Process change	Company
18. Data clean-up	Company
19. Data conversion	Vendor
20. ERP issues identification	Company
21. Scope change	Company
22. Software changes (coding)	ERP vendor
23. Management action to issues	Company
24. Go-live readiness reviews	Company/vendor
25. Live running cutover	Company/vendor
26. Post-ERP cutover	Company

On looking at the areas for management of the project, the responsibility, and the level of effectiveness, it became clear that no one in the organisation has been given a mandate to effectively manage the amount of issues arising in each area. Even when organisations hire external consultants or software consultants to project-manage their ERP project, they typically focus on the technology aspect of the project and are not given the authority to look at all of these areas listed above.

There is an assumption that these issues have been dealt with when in reality, some may have been looked at whilst others have not even identified as issues in pre-project planning. The way that these activities are or are not handled prior to software even being purchased can have a major impact on any ultimate ERP outcome.

Many downstream ERP project problems can be predicted long before the project implementation stage has even begun, which would provide the organisation the opportunity to prevent the problems from manifesting into disasters.

A prime example is the 'acquisition rationale'. If senior executives have not considered all the issues surrounding the acquisition of an ERP system with the detailed expected outcomes, then how can they clearly articulate instructions on budgets, resourcing, live running, etc., to the project management team and employees in the company who will be tasked to undertake the work and implement the system?

If you don't know where you are going, any road will get you there!

Why do some companies achieve a better result than others? Because they manage the entire spectrum of the project better and don't just focus on the obvious technology issues associated with software.

For many organisations, the fate of the ERP project is sealed long before the purchase of software or the project implementation phase begins. Disasters don't manifest themselves at the beginning of projects, and projects often proceed in blissful ignorance of the looming disaster that is a short distance down the line.

The problem is most companies do not see the ERP project as a company change project but instead see it as a technology project that can be handed off to others to undertake. Wrong!

An In-depth Look at the Issues

1. ERP Acquisition Rationale

ERP systems typically represent a significant change to the way we look at and strategically run the business. The real-time integrated nature of the software tools can provide a significant competitive advantage over traditional ways of operating. Senior management's understanding of the total capability of an ERP System, at the beginning of the ERP acquisition, is critical to the way the project is approached and communicated to the rest of the company.

The rationale for adopting or changing to a new ERP system, from a strategic perspective, can impact the outcome and can result in a disastrous failure. Management needs to understand the issues involved in a successful ERP project and take a broader view than just a technology project. In a manufacturing organisation, this would include understanding the ERP operating philosophy to obtain the optimum benefits.

An example of owning versus using is to buy an aircraft. Owning the aircraft doesn't mean you can fly it! You have to learn how to fly it before it is of any use.

Some of the issues to be considered as part of the acquisition rationale would include the following:

- Why do we need an ERP system or upgrade?
- What detailed outcomes can we expect by department? State KPI improvements.
- What amount of work across the company needs to be undertaken to make the project a success?
- How will we communicate this to the organisation to get everyone on board?

2. In-House Expertise in ERP

ERP systems and their former MRPII systems use a complex collection of software modules that must be run in a certain way as a philosophy for running the business. That typically requires a significant change in the management of information and the decision-making process within the company, both management and employees, to make it work properly. An ERP system can be successfully implemented in terms of software and data but fail due to the way in which we make decisions and the processes we employ to do things.

An understanding of the ERP philosophy, process re-engineering, how to apply the software, and what needs to change at the detailed level in the organisation are essential components of success and can be learned through an education process on the use and implementation of the technology. Using an ERP system simply to run things in the same way as before implementation is a guarantee of disappointment.

The question of relevance on in-house expertise is, what level of expertise does the company have in implementing and operating ERP? It is easy to gloss over the issue of expertise as you may have a number of employees who have participated in an ERP project previously, but what their level of involvement and how successful the implementation was are key questions to be answered.

Embarking on an ERP project with limited knowledge simply puts the organisation at the mercy of ERP vendors and their implementation partners, whose primary interest is in extracting the maximum amount of revenue, irrespective of the ultimate outcome.

Independent education on ERP and the issues involved can be beneficial to senior management in their decision-making and understanding of the issues involved in any ERP project. This is a step that most organisations fail to consider preferring to delegate the detail to others in the organisation.

The total spend on education (as opposed to training) will always be a small fraction (<5%) of the total cost of an ERP project. It should be regarded as really cost-effective insurance given the potentially huge costs involved in the whole project!

Given the large number of poor ERP outcomes, it is vital that CEOs and their senior management team get as much information as possible prior to committing to an ERP project.

3. ERP Risk Assessment

ERP systems provide organisations with a potentially powerful tool that can strategically change the way the business is run and herald real-time integration of all the functions within the business. Whilst there is potentially a very big upside, the downside poses significant risks if it goes wrong. A full risk assessment prior to the project taking life can significantly mitigate the risk and enable effective plans to be made, if required, to meet the risk. The risks that need to be covered include the following:

- Financial risk
- Technical risk
- Project risk
- Political risk
- Cultural risks
- Business disruption risk
- Contingency risk
- Software failure risk
- Non-use, misuse risk
- External risk
- Competitive risk
- Reputation risk

For many companies, risk analysis is too simplistic, and they cover some of the obvious issues whilst ignoring others that may come back late to bite them. The statistics on ERP failure are real, and effective risk analysis is essential.

Senior executives should not simply assume it has been done correctly!

4. ERP Cost Justification and Budget

Given the potential for running into trouble with an ERP project, it is vital to state what the ERP expectations are in detail and justify how those returns are going to be achieved. Setting a budget based on a detailed assessment of the work to be completed and the costs involved instead of just assuming, without knowledge, of what is involved is a platform for future cost and time overruns on the project. A proper cost justification that covers the actual work, hardware and software, to be undertaken and contingencies that may arise balanced off by the expected benefits the technology will provide will give a realistic expectation of cost, times, and returns.

Typical areas to be considered:

- In-house hosted versus SaaS (Software as a Service) or cloud hosting
- Software specification/model development costs
- Software evaluation costs
- Software licence fees
- Hardware costs
- Software maintenance costs
- Software integration with legacy systems
- Hosting costs if SaaS
- Education costs
- Training costs
- Consulting costs
- In-house implementation team costs
- Data clean-up costs
- Process re-engineering costs
- Business disruption costs

The costs to be balanced off by the expected gains and returns to the business would include the following:

- Productivity increases (but how do you measure this!)
- Headcount reduction
- Process improvement benefits
- Inventory reductions
- Customer service improvements
- Quality improvements
- Increased sales
- Reduced scrap
- Bottom line improvements

Whilst these overall categories may be some of the areas where costs and savings will be incurred, it is important that any savings are realistic and not just vague statements, unquantified, for some gains somewhere in the future.

Where savings have been identified as part of the cost justification from the different areas of the business management should sign-off as their commitment to the ERP project.

Note that it may be the case where the acquisition rationale is driven by a non-quantifiable measure such as 'we need an ERP system to improve our competitiveness in our industry'. If that is the case, then the justification document should be built around this reason and not around a bunch of fudged, vaguely quantifiable KPI improvements constructed just to please the CFO!

5. Request for Proposal

Understanding the ERP philosophy and the issues with failed implementations provides insight into the information we put into a request for proposal from different software providers. The proposal *must* contain a visual model as to what the software needs to do with specific instructions on functionality, hardware platforms, number of users, transactions, etc.

The importance of a model cannot be overstated! *(Refer Fig. 2.)* Having a model against which to test any proposed ERP software ensures software functionality fit *prior* to buying the software.

A major complaint from failed ERP users is that elements the software were inadequate or not suitable for use in the business. Discovering software inadequacies during the implementation stage of the project or after live running is achieved leads to purchase of additional modules or modification costs, sometimes large, delays in the project, or chaos after the system is turned on to live running.

Developing a model together with how you wish to process through each stage of the model, including a number of expected transactions, format of product bills of materials, forecasting, and any other special requirements, enables a determination of the effectiveness and fit to the company requirements of the software prior to agreeing to buy the software. Problems experienced during simulation of the system, through the model, is a good indication of the problems a company would experience if that software was purchased.

ERP software vendors would prefer and resist going to the trouble of demonstrating their product against a model, and some may even suggest that setting up a simulation with your data is too difficult and cost prohibitive. This indicates they do not want to expose their software to the shortcomings they would experience during the simulation. Part of the reason they don't want to undertake the work upfront is that they like to charge you for modelling and software configuration, post purchase. By developing your model upfront, you have already determined what the configuration will be and will have tested it to ensure it hangs together.

Make it a precondition in your request for proposal that they must successfully simulate their software through the model, using your data, in order to be considered as a supplier. They may argue that there is a cost involved in this and may request some upfront payment. This payment is worthwhile considering, but negotiate and advise them that they must firstly agree that the software meets the requirement, in writing. Money spent here can save you money in ensuring the product fits the business requirements.

If they are not prepared to agree to this, simply walk away and move to vendors that agree to the simulation.

The request for proposal should be specific but open to alternatives as new innovations may be of interest as an organisation wouldn't want to exclude possibilities of advancements in the technology.

It can be difficult to interpret responses from potential vendors, and it is worthwhile phrasing the requirements as a series of questions that require a yes, no, or would require modification, which makes it easier to cull out those packages that do not meet the requirements.

The development of a model that fits the operating requirements of the company and especially specific operating requirements is a critical step in any ERP project. Failure to adequately define the model can only result in downstream issues requiring costly modifications and delays to the ERP project during the implementation stage of the software component of the project.

Be careful not to be bamboozled by jargon designed to obscure deficient software functionality. They either have the functionality that can be seen or they don't or it would require modification to achieve the desired functional result.

Be wary of vendors who use terms such as 'best of breed' or 'best practice software for a specific industry' as this is nothing but sales hype that can give a false sense of software suitability for the organisation only to find during the software implementation stage that it has problems that are costly to resolve.

Testing the software through the model is the best way to eliminate the issues of fit for purpose and give the buying company confidence in the product selected.

The issue of developing a model to test and simulate the operations of the company through the software is one of the major failings of companies embarking on an ERP project. Surveys showing suitable software selection as being a major issue causing scope creep and budget blowouts can simply be addressed by taking the time to construct a model with the detail of what each function requires within the software.

Models can also be used to identify areas of non-performing ERP systems by simply colour-coding the model to reflect the operational status.

Refer to the following macro-model example:

Fig. 2: Sample macro-model (note each company must design their own model to reflect their own business)

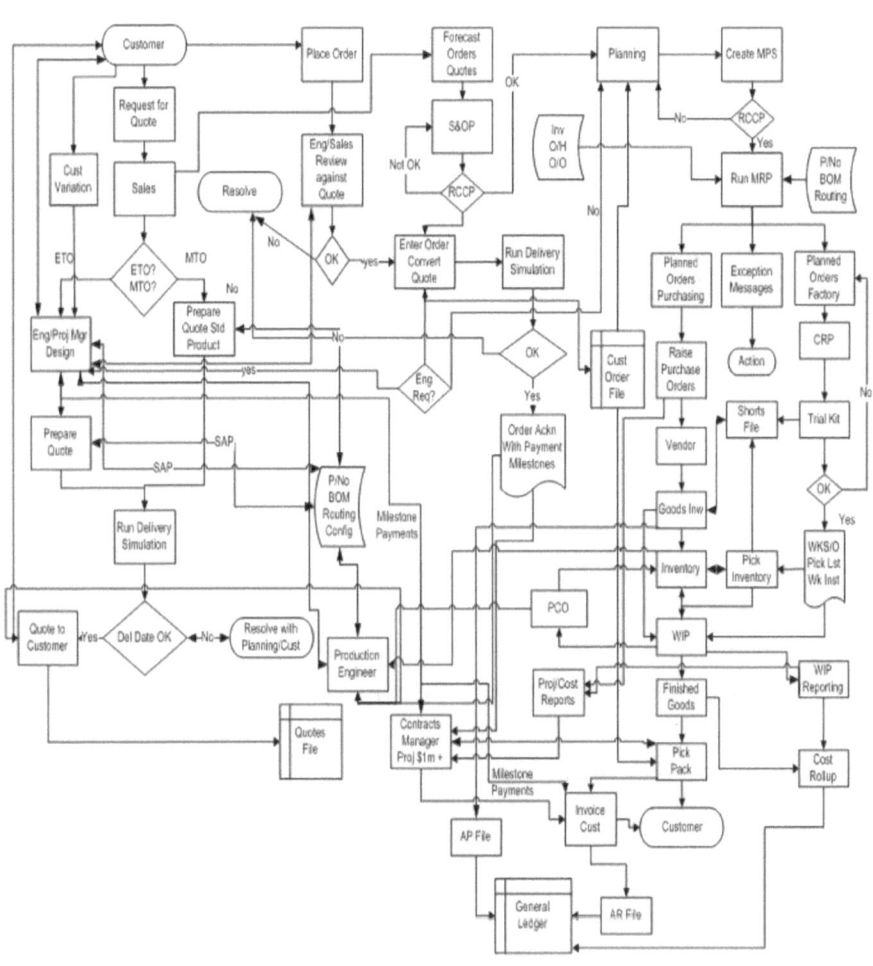

Whilst the model may look complex, the reality is that it reflects how the organisation must process using the technology. The risk is that the software will not be suitable if you cannot simulate your operation through the software.

6. Software Selection Criteria

The selection of the ERP package is a key factor in cost blowouts as during the implementation stage, it is discovered that the software, whilst satisfying the functionality criteria, may be unsuitable for application into the process for your specific company.

Testing the software through the model and your specific requirements ensures that the functionality is suitable for your organisation.

An example of this is a company that specified the software must be able to forecast across a number of algorithms, and whilst the software had this capability, it could not break the products into families and distribution points. A classic case of insufficient information at the specification stage!

Some issues to be considered when evaluating potential ERP software packages

A weighting measure needs to be assigned to each module of the software so a decision can be made based on the needs of all areas of the business being met and not just mandated based on narrow departmental such as accounting or computer department biases

Issues to be considered:

- Hosting SaaS or cloud versus in-house hosting
- Web access and integration
- Reporting and analysis capability
- Administration and security

- Functionality fit
- Systems architecture, including integration with other systems
- Software cost:
- Number of users
- Configuration costs
- Consultancy costs and guarantees
- Potential future costs
- Maintenance cost
- Training costs
- Customisation costs
- Help desk costs and response time
- Software company financial stability
- Length software house and product being in business
- Date product first released into market
- Software market penetration
- Software company reputation
- Reference checks for company and proposed consulting staff CVs
- Frequency of revision releases
- Complaints about ERP vendor from other industries

Where ERP vendors are engaged to undertake implementation consulting and project management activities it prudent to examine carefully what they are offering and decide whether you really need these services and if so ensure any payments are negotiated based upon specific milestone achievements

Where fixed price contracts are negotiated having the milestone specifics included are a way to contain costs of vendor add on consulting services control any escalation of costs based on project delays.

The cost of these implementation services tends to be disproportionate to the expertise required or provided. Many ERP vendors, their implementation team or partners, cover this up with unnecessarily complex implementation plans developed with their revenue stream more in mind then a successful ERP project outcome.

An alternative to vendor implementation services is to establish in-house implementation teams and a senior project manager supported by an independent consultant from one of the small boutique specialist ERP consultancies who visit on an ad hoc basis, leaving the ownership of the project in the hands of the company. This seems to provide much better cost-effective outcomes, provided all the other steps are followed.

It is worth bearing in mind that at the upfront negotiation stage for software and services, a lot of goodwill exists on both sides. The problems arise when the buying company proceeds into the implementation stage, where the project starts hitting roadblocks.

Most ERP projects' success or failure is determined well before the acquisition of software simply by the approach taken to the selection of ERP software and the project overall by the company.

7. The ERP Contract

Many complaints are made about the performance of the software vendor and their implementation team with regard to software functionality, consultant activities, deliverables, and project timing.

The most significant issue with a lot of ERP contracts is that they are not specific in terms of what the software house is responsible for and what the company is responsible for.

The goodwill generated at the beginning of all ERP implementations quickly evaporates into disillusionment with progress due to both the company not providing resources or information and the implementation team being frustrated with the lack of focus on the part of the company.

Whilst it is never possible to cover every contingency in the contact, it is possible to define the deliverables and the timing to be delivered. The model and associated agreed operating requirements should be incorporated into the contract with specific objectives stated.

ERP software vendors and their implementation teams/partners do not like being held accountable for outcomes despite charging huge sums of money for their products and services. Payments should be linked to milestones being successfully achieved and a penalty imposed if these milestones are not met.

ERP vendors and their implementation teams/partners do not like the concept of accountability and will go to great lengths to avoid accepting responsibility for work they undertake. This is despite the heavy sales claims about products being 'best of breed' and 'proven path' implementation methodologies.

In-house issue escalation clauses can and should be written into contracts that provide all parties with the mechanism to flag problems early on to senior management so they can be resolved in a timely manner and not evolve into a crisis.

The contract should include a nominated neutral third-party mediator, who will become involved in mediation, in the event of a dispute, and whose decision all parties will agree to abide by. Again, ERP vendors and their implementation teams/partners are reluctant to agree to any clauses that could hold them responsible for issues arising from their project activities.

Carefully worded contracts that protect all parties from the negligence of the other parties can be framed in the interests of all involved. The reluctance or refusal of ERP vendors and their partners to frame the contracts in this way and to accept accountability is an indication that their sales hype of 'proven path' implementation methodologies is little more than spin.

There seems to be considerable evidence to suggest that not hiring the ERP vendor's implementation team/partners to project-manage and provide services and instead looking to in-house expertise and ad hoc independent consultants may be in the best interest of the ERP-buying company.

Disconnecting the software sales process from the implementation process may be a better approach as third-party implementation consultants have no vested interest in covering up flaws in software functionality and are not looking to expand the scope of the project.

The differentiation of responsibility should be clear-cut. The ERP vendor and its team should provide software, configuration, training, and data conversion, and the project management should be either an in-house project manager with ad hoc assistance from a third-party ERP independent implementation specialist or a third-party project manager supported by in-house project teams.

8. Project Plan

The ERP project plan must be formulated to include all the 26 steps that can impact a project. There may be other considerations that individual organisations need to include. The project plan goes well beyond the technology software component and must be approved by the CEO of the ERP-buying company.

An effective ERP implementation plan should be no different to the detail required for any major project . . . *It is a major project!* It must be comprehensive and encompass all the activities required for a successful outcome. Focussing too much on the technology component without consideration of other aspects of the technology will only lead to poor outcomes.

ERP software vendors like to present complicated, jargon-filled plans designed more to bolster their own revenue streams by trying to reflect a complexity that justifies their own involvement rather than a comprehensive, easily understood plan that can be followed by the company in implementing the technology.

The ERP plan must include the technology component in sufficient detail to enable it to be followed and reported on.

Regular reviews must be included in the plan to ensure the business focuses on the implementation and can react in a timely manner when issues arise that require management's attention.

Specific measurable milestones must be built into the project plan that are tied to the payment schedule for the software and any services the software house provides. These milestones must be relevant and easily tracked.

The monitoring of the plan is a critical function that the project manager does, which also involves the senior executive and his oversight, to ensure the company is in control of the project and will not let it slip and simply run out of control.

It is worth the time to spend the effort in detailed planning for the project and debating the plan before the implementation effort commences. The plan should be comprehensive and cover off all the 26 points.

9. Go-Live Date(s)

The trap a lot of organisations fall into with their ERP project is to mandate a date for live running based upon an event unrelated to the amount of work that has to be carried out to successfully complete all the necessary implementation activities, such as beginning of new financial year.

A successful ERP go-live event should be based on a realistic plan for the completion of work such as data clean-up, training, process redesign, procedures, system testing, issues resolution, and overall company preparedness for the live running.

Setting a go-live date based upon a non-related event such as new financial year and not on the actual amount of work that has to be done to achieve live running will inevitably cause short cuts to be taken which will contribute to the system not performing and chaos when go-live status has been achieved.

The money and time organisations think they are saving by taking short cuts will be more than exceeded in operating problems and losses when the system is turned on.

Taking the time to do the research into all the activities required to implement the ERP system will ensure a more realistic understanding of the work, cost, and time involved in the ERP implementation process. This must be done at the beginning of the ERP acquisition consideration prior to committing the organisation to the project.

10. ERP Education

Understanding the philosophy behind ERP is fundamental to achieving the right result from the technology.

Understanding how to run the business, in the right sequence, to obtain the maximum benefit from the technology seems to be missing in the vocabulary of the majority of implementation plans.

Education is needed for the entire company from senior management down to the people at the coal face so everyone understands what the technology is, the integrated nature of the technology, and the impact of actions in one area on other areas of the business.

The genesis of ERP was back in the late 1960s in the form of MRP, which progressed to MRPII and then ERP. The fundamentals of materials requirement planning (MRP) still reside in ERP systems, and the issues today are exactly the same. Data accuracy on product definitions, parts master files, bills of materials, inventory accuracy, lead times, scheduling, factory planning and feedback, customer requirements, etc., are fundamental to the operations of the supply chain for the organisation, and the post-live running implications of incomplete and incorrect data can affect the entire supply chain and cost organisations millions of dollars.

The education program should cover each of these areas, relevant to the type of business to ensure the entire organisation understands how the philosophy works, options in structuring data, and what has to be done to operate effectively.

In ERP implementation projects, we see the words 'training' of users, and frequently we see missing the words 'education' of users.

There is a distinct difference between training and education!

Training is the 'how to use the tools' (which buttons to press) to perform a task. Educating is explaining the reasons 'why' the task is performed and the bigger picture philosophy of how to operate using the ERP tools and its role in integrating the organisation's business processes. An example of the differences between the issue of education and software training is analogous to training someone to fly a jet fighter but not providing education theory on aerial combat, the principles of flight, or concepts of jet propulsion.

An effective education process gives people the knowledge to know what, when, and why to do things and not just know how to press button to perform software functions.

11. ERP Training

ERP user training is a large source of complaint for many organisations. ERP software providers, typically, prefer to get the training component of their project out of the way, which has two implications. Firstly, training held too early in the ERP project cycles tends to be forgotten and leaves a huge gap in knowledge when the system goes live, and secondly, the ERP software provider sees additional training as a source of additional revenue from the implementing company.

The quality of training can also be a source of contention as many companies complain the trainer did not have sufficient knowledge of the system and when questions are posed by the people being trained respond with 'I'll have to get back to you'. In my experience, they rarely get back to the questioner with an answer.

Training should be scheduled at the optimum time to suit the company and not the ERP software supplier!

A very popular strategy is for 'super users' to be identified in the company who are trained in particular aspects of the systems operations who in turn trains the other users and becomes the 'expert' in that part of the system. In theory, this sounds OK, but in practice, the super user ends up being the person everyone sees as responsible for the function, and others either don't attend the training or don't take the training seriously as they believe the super user will solve all the problems that arise.

The concept of super user is fine in theory, but in many cases, super users become frustrated due to expectations of them and end up either not performing the function properly or quitting the company.

A training matrix should be created for every employee with the areas they need to be trained in, the date planned for the training, the date training carried out, and whether further training is required. Training should be scrutinised by management to ensure employees attend the training sessions and follow-up with ongoing exercises so that the training is not simply forgotten.

12. Implementation Responsibility

The concept of the 'Devil's Triangle' is one coined to reflect the implementation players in an ERP project. There is the ERP software vendor, the vendors implementation team (sometimes a third-party vendor partner), and the company. Each of these players has an agenda that is typically in conflict with the other players. Their agendas in a nutshell are as follows:

- The ERP software vendor's agenda is to create a vision through selling features and benefits of a package that are rarely, if ever, achieved by organisations that have bought the package.
- The ERP vendor's implementation team/partners, who try to implement the vision sold to the client knowing that the package has been oversold or even misrepresented, feel safe in the knowledge the client will fail to meet deadlines and ask for modifications, which will extend the project timeline, increase the revenue for the vendor's implementation team/partners, and provide the basis for the reasons for the project failure. These vendor implementation consultants focus on the technology aspect of the project and do not get involved in the corporate areas of responsibility.
- The client who has been sold the vision but ends up wearing the results of the accumulation of costs and issues resolved and unresolved during and at the end of the project.

Typically, implementation plans are overly complex in order to justify the huge implementation fees that are charged for the service – the plan that is supposedly a 'proven path' or other words that are designed to give confidence to the buying client that the implementation team knows what they are doing. Often, these plans are not followed and quickly become redundant in the overall project scheme.

With up to 70% of ERP implementations not achieving the outcomes expected, it would be fair to say the vendor's implementation team are more experienced in failures than in success.

The responsibility for the ERP project with these three players committing resources to the project can lend itself to confusion as to who is actually directing the project activities. The ERP vendor and his implementation partners rely on the client's ignorance of ERP and plan how it is to be used to dominate the implementation process and increase their own revenue streams whilst blaming the company for missed deadlines.

The ERP project plan should contain all of the 26 steps to ensure they have been adequately covered. A number of the 26 steps are not software related and typically tend to be overlooked and are certainly not within the mandate or expertise of third party consultants to deal with them. Senior management should ensure the project plan has sufficient management oversight to identify and deal with any issues that have the potential for causing problems during the project.

For the majority of ERP projects, failure has been determined long before the actual implementation of the technology aspect of the ERP project has commenced.

Senior executives' understanding of the entire scope of an effective ERP project can put the company in a position to deal successfully with the issues from a perspective of understanding instead of dealing with the ongoing crisis of an implementation project out of control.

The ongoing oversight of the project by the most senior executives is essential to ensure the company stays focussed on the ultimate project success!

13. Project Management

To be successful, the project management for an ERP project must start long before software is purchased. The 26 steps represent the issues that an organisation must address for the project to be effectively managed and controlled through the entire process. Clearly, these steps represent far more than simply the technology implementation of a computer system.

The project management of the entire ERP project should be seen as a 'business transformation project' with the senior management in the company keeping a watching brief to ensure all these steps are covered.

The project manager appointed by the company must be a senior person who has the capability to understand the quantum of work and has access to and the confidence of the CEO to push the project through to a successful conclusion. I am often asked who the project manager should be. I find someone from the operational side of the business tends to make better project manager as they are focussed on getting the system right for the operational side of the business. IT and accounting tend to focus on either the computer system or accounting to the exclusion of the other parts of the business. Having said that, I have seen IT people do an excellent job of project-managing ERP projects.

The project manager must be a good communicator and be able to formulate and manage the details of the project plan. The project manager should be seen as a core manager within the business and not just as a project manager who is assigned to some peripheral project that senior management can ignore.

The project manager should be supported by a cross functional team(s) from each of the different operational departments who will be responsible for carrying out the work to implement a successful ERP project.

The project plan is the blueprint for the project and it must be updated to reflect any additions and project progress and be reported on to management at least monthly.

Where the ERP software vendor's people are assigned to the project, the project manager must ensure the company is getting value for money and not just have the consultants simply booking hours to the project on activities that add no value.

The project manager should be in constant contact with the CEO to ensure any issues beyond his authority are quickly recognised and resolved.

14. Executive Involvement

Since the early days of MRP and MRPII and now ERP systems, it has been identified that senior executive involvement is an essential ingredient for successful ERP outcomes, and indeed, it is a major issue regularly cited in ERP failures.

A major problem is that senior executives do not see themselves as sufficiently competent with the technical aspects of ERP systems and instead delegate the responsibility to others they feel are more competent to undertake the work. The problem with this approach is that only limited authority is granted to those delegated to implement the system and that no mandate is granted to cross all the inter-departmental boundaries to make the changes necessary for a successful outcome.

The senior executive involvement need not involve technical knowledge of ERP systems but does require an oversight of the project activities, an understanding of the potential problems being faced, and a willingness to arbitrate issues that the project manager is not able to resolve.

Education for senior executives and management provides a level of understanding that is essential to support the ongoing ERP effort. Many senior executives and management believe that education is for the system users and that it is not necessary for them to become involved in the education process. This is a major mistake, and the lack of understanding of ERP leads to lack of support on ERP critical issues impacting outcomes.

The project plan should be a priority of senior executives, and this plan should be constantly monitored to ensure the project is on track and on budget.

Senior management must establish an environment whereby the project manager and project team can raise issues safely in the knowledge they will not be criticised, ignored, or simply put into the too hard basket. Ongoing communication from the senior executives to the entire company provides a positive, supported platform for the ERP project.

An ERP project will require changes across departmental barriers, which can cause significant internal disruption and conflict if not managed effectively. A strong executive team, educated in ERP and driven by top management with support that is more than just lip service, will be the difference between success and failure for the ERP project.

Lack of senior executive involvement is a real issue and must be addressed from the beginning of an ERP project from initial discussions to final implementation and beyond!

15. Software House Expertise

The ERP industry has not exactly covered itself in glory over the past 20 years. The number of ERP clients who feel that they have been fraudulently misled by software suppliers and their implementation partners is on the rise, as is the tendency for unhappy clients to seek legal redress.

Some of the more odious gouging activities of ERP software vendors include activities worth noting such as the following:

Bait and switch. This is the practice of displaying certain consultants, during the sales process, to show the sales company understands business and the ERP implementation process to ensure a successful outcome. These consultants understand the software and may well understand business issues that will come up as part of the implementation process. The problem is when the client buys the software with the expectation that these are the people who will support them in the implementation process only to find they have been allocated people who have lesser skills and in fact may be on a learning curve using the client as the training ground.

Resumes: lies and misleading half-truths. Resumes are written to deliberately mislead the client into believing the persons proposed have the skills necessary to support the implementation of the system. This is a common practice as the resumes are more like a vague overview of the skills and experience of the person being proposed. Many of these so-called consultants lack any sort of business skills but may have some knowledge of the software sufficient that the implementing client believes they are well served until the project starts experiencing problems. These consultants are very good at identifying where the project has gone wrong but only after the project has hit problems. They are also skilled in apportioning the blame to the client to absolve themselves or the ERP software supplier or partners of any responsibility.

In many cases, consultants are advertised for when an ERP software vendor has obtained a new client and the software house needs to get manpower to staff the project. The consultants employed in this way may have minimal knowledge of the software or its application into the target company but are able to obscure their lack of knowledge through the use of jargon beyond the understanding of the target company.

Intentionally under-quoting. Software houses are usually aware that the prospective client has a budget that they will evaluate and make a decision on. History shows us that the cost of implementation, if using third parties, is many times the cost of the software. This is understated during the negotiation stage with the ERP vendor knowing, from experience, that the project budget will blow out during the implementation process and that the blame can be sheeted back to the client.

Consultants are aware that changes will be requested during the life of the project which provides an ideal platform for gouging the client where the client has little choice but pay up or risk the investment already committed to the project. This is no difference from any major project – low quoting to win the job and then raising additional revenue with expensive project variations, modifications and add-ons and being able to charge any costs and delays back to the client. The impact of this can be minimised, if not eliminated, by constructing and testing the software against a model and operating criteria and then agreeing this as the basis for the contract supply with the supplier and tying this into the contract and contract milestones.

The 'proven path' implementation methodology. ERP-prospective clients want assurances that the company they buy the software from knows what they are doing and can provide support when required during the implementation stage of the project. All software vendors see the implementation of ERP as a cash cow that has evolved over the past 25 years or so.

It is a very attractive proposition for CEOs and boards to buy an ERP package and implementation services that are titled proven path believing that they will have a successful outcome with the project. With the failure rate from 50% to 70%, depending on the criteria for failure, the term 'proven path' seems to reflect the proven method of extracting large amounts of dollars from hapless companies and then not delivering satisfactory ERP outcomes.

Often, sample plans are put forward at the sales stage that are quite extensive and filled with jargon to justify the enormous dollars asked for the service, and when they commence the implementation process, they use other tools that are not even in the sample plans and so-called proven path. It is important to note that the definition of success usually has two translations. The software house translation is the software is running, and therefore, it is a success. From the perspective of the implementing company, they want the software to run the business and get the results they expected or cost-justified for. Success on one hand can be a failure on the other!

It is safe to say that the 'proven path' vocabulary being used in the sales cycle is simply to mislead the buyer and close the sale.

Ignorance is gold. ERP salesmen love visiting a client who really has little idea what an ERP project is about and who looks to the integrity of the selling organisation to be honest and fair.

Most potential ERP buyers that have no model, ill or poorly defined project outcomes, an undefined scope, or lack of basic education of ERP are seen as a gold mine. The vendor's sales strategy is to gloss over critical details and issues until after the client signs the contract. Average budget overruns are in the region of 25% over budgeted costs. Some implementation cost overruns can cripple companies financially or bankrupt them!

The reference sites. When a potential client wants to perform reference checks, many ERP vendors only provide the names of a few reference sites. Interestingly, they claim hundreds of successful implementations but only offer a few sites we refer to as 'marquee sites'. These sites are often compensated by the ERP vendor in some form for being a reference site.

Don't expect these sites to provide any independent perspective on the software package or the vendor or any negative news about the vendor!

It is worthwhile to ring the reference site and speak to the operations people, not the contact given, and question them about different aspects of the systems operations and how the implementation proceeded. The ERP vendor does not like this as it exposes the truth and may jeopardise their potential sales to the buying company.

Milking the cash cow. Once the implementation process starts and the consultants are in the company, their instructions are to try to sell their additional services and book hours. These include more consultants if the project is slipping, readiness assessments, cultural change programs, and so-called best practices, to add functionality that was not included as part of the original software package, sometimes deliberately, to keep the software cost within the budget even though the salesmen knew it was required to make the system work.

There is also a tendency for ERP vendors to offer inducements by way of discounts for software modules that the client doesn't need but might later on. This is a total waste of money for the client, but these people are very persuasive. Use the model to determine what software modules are required as it is difficult to persuade companies for additional software modules if the modules are not required to support the model.

Few issues cannot be resolved by educated company people who understand what ERP is and is not and are able to understand that implementation is really about project management of configuration, process re-engineering, cultural change, management involvement, data clean-up, training, procedures, and testing of systems to ensure they work as they should before turning the system on. The complexity myth that has evolved over the last 20 years or so has been perpetrated to ensure the continuity of the ERP industry gravy train and is not to the benefit of the ultimate customer of the ERP system.

The promise of knowledge transfer. A great confidence building measure during the sales stage of the ERP system is for the potential ERP vendor to use words like partnering and the promise of knowledge transfer to the client by using the consultants from the software house to implement the system.

The software vendor or his implementation partner will state it as an objective of transferring software and ERP knowledge to their client. Mostly, however, this doesn't happen, and the client has to force the issue and ask them to leave. If you look at the hourly rates being charged for services, what incentives do consultants have to transfer software or ERP knowledge to the client, if the client continues to pay?

The ERP industry is worth around 50 billion dollars, and with so much money literally being thrown about, fraudulent and unethical practices are inevitable. The best defence against this is to seek independent consulting advice from smaller boutique organisations that have no vested interest in software and have a verifiable track record. In any event, in-house implementation control using your own people with ad hoc consulting assistance is a cost-effective way to manage an ERP project, which, after all, is about 'project management', a technique used to manage any project of any size.

16. Software Implementation Team Expertise

The questionable practice of ERP software vendors and their partners putting inexperienced people into client organisations and charging exorbitant rates for expertise the people simply do not have is common in the ERP industry. The best way to insure against this happening, if indeed you need to engage the vendor's consultants, is to insist on a vetting practice whereby references are thoroughly checked and the proposed consultant(s) has to be approved by the company for specific tasks before being permitted on site.

The company should be able to contractually remove consultants where they show a lack of expertise or do not perform functions to plan. The removed consultant should be replaced, if required, with an experienced consultant after the vetting process confirms their suitability.

Consultants will always generate an air of being extremely busy on project activities as this justifies their being on site. To ensure you are getting the most from experienced consultants, have a detailed plan of the activities they are undertaking and a time for the activities to be completed. Do not leave it up to the consultants to determine their own activities as they will always justify their time by being involved in issues where they are not required. Ensure milestones are set for the consultant, the tasks are competently carried out, and the consultant is then removed from the site. Otherwise, they will simply stay on site charging hourly rates for work that is not necessary for them to do. Their companies charge them out at an hourly rate. The more the hours, the more the revenue for the contracting company!

17. Process Change

Business processes have evolved over many years and become an integral part of an organisations culture. With the advent of ERP type systems, processes can and should be re-examined and altered to take into account the automation and capability of the tools within the ERP system.

The term 'process re-engineering' has been used to describe these changes. Process re-engineering can take place without an ERP system, but redesigning processes around ERP software capability ensures that business processes are matched up with the ERP system. Failure to match the process with the ERP system can derail the ERP project and create chaos when the ERP system is turned on.

Changing processes that have evolved over many years inevitably touches on the culture of the company and creates resistance to the changes and if not managed carefully can undermine the benefits of the ERP project.

To overcome this resistance, top management must provide leadership and oversight for all proposed changes and overcome objections and disagreements that arise in the process of re-engineering and ERP implementation.

Departmental managers will put barriers in the way of changes if they perceive that their area of influence is diminished or they feel uncomfortable with the change because it is change.

The other significant issue to deal with is the mandate given to the ERP project manager to cut across departmental boundaries to initiate the changes required to get the maximum advantage from ERP. This can be a major factor in the progress and outcome of an ERP project. Ensure you have a process in place to justify changes and the changes can be explained logically in order to gain acceptance.

Many organisations find the internal disruption and cultural issues too difficult to manage during the ERP implementation and opt to put the re-engineering off until after the ERP project has been completed. This is problematic as decisions need to be made concerning the software configuration and processes during the ERP implementation and organisations rarely revisit the issues due to complexity and ERP corporate exhaustion. (Everyone is fed up with ERP and is not interested in revisiting it with process re-engineering.)

Process re-engineering can significantly improve the operations of the business by modernisation, elimination, or reductions in cost-adding activities. An ERP project is an excellent platform for making these changes, but it will not happen without management driving it.

18. Data Clean-up

Establishing accurate and consistent data is essential for supporting the supply chain and realising significant returns for the ERP software investment and achieving improved business processing outcomes.

A common issue with many organisations in implementing an ERP system is how they are going to be able to extract their existing data from their old systems, clean it up, and transfer the data into their new software system. There are many instances where existing data is duplicated and contains inconsistent information. Examples of these are in legacy systems where products can be entered using a product description and called something else in another part of the business or system. In many instances, the data has been poorly maintained and contains significant errors.

The software implementation process is an excellent opportunity and an essential one for the organisation to clean up its duplicated, incorrect, and inconsistent data and develop a single data set for products, customers, ingredients and raw materials, suppliers, financial accounts, etc.

While the process of going through potentially hundreds or thousands of data records may be painful, it will prove to be highly beneficial during the initial system go-live and beyond.

Data clean-up is the source of potential future problems as companies finding themselves overrunning budgets and time milestones attempt to bring their project back on budget and time by taking short cuts in data clean-up and migration activities only to find that the cost and disruption of rectification, post-live running, are disastrous for the company and costs many more times than it would have if done during the implementation stage of the project.

Short cuts during the implementation process that might make sense at the time can cause millions of dollars in losses in business disruption, chaos, and in some cases send organisations bankrupt after the system goes live.

19. Data Conversion

ERP software systems process data! If inaccurate data is transferred into the new ERP system, the system will process it, but the output will be incorrect, causing confusion and chaos.

Converting data from legacy system to the new ERP software should only be undertaken with a safety net, namely a well thought-out plan of execution. ERP vendors (theoretically) have data conversion software that is able to be configured to take an organisation's data and convert it into the form they require in the new ERP system. A word of warning here! Data conversion has been the source of major delays and disputes in ERP implementations and must be approached in a manner that ensures a clean migration from one system to the other.

The issue of when you should manually enter the data and when you should electronically convert it is related to the specific file type, whether it be inventory on hand, quantities, customer account receivable balances, or sales history. Whether talking about a single application or an ERP software package, converting legacy data is an issue that merits discussion and careful planning.

Obviously, the conversion considerations for an ERP package, with all of its touch points, are far more involved and could require significant technical resources. Furthermore, since the effects of the conversion methodology may impact the end-user, these users should be involved in the discussions and decisions.

Data conversion is a critical step in the implementation of ERP software. As such, it deserves careful and important attention and should not be treated as an afterthought or over-simplified!

20. Issues Identification

During the process of the ERP project, many issues will be identified which will be examined, researched, and resolved as part of the process and within the skill and mandate of the project manager and the project team. There will be issues that are identified outside the authority and skill set of the project manager and the project team to deal with that could cause significant problems if not resolved. This has proved to be the source of many complaints over the past 35 years! That is, the lack of senior executive involvement in resolving problems that are beyond the mandate of the people working on the implementation of the system.

Many senior executives take the view that ERP is a technology project and that the detail should be left to the IT and computer people. The problem is there are a lot of non-computer issues that go with an ERP project that can only be resolved with the input, involvement, and ultimate decisions of the senior executives of the company and in particular the CEO. ERP issues will cross departmental boundaries and can run into stiff resistance from managers who may see their own area of influence under threat, and they can derail a project unless direct action is taken by senior executives to ensure this does not happen.

To ensure senior executives and the CEO are kept in the loop and are able to intervene in a timely manner to resolve difficult problems and disputes, an 'issues log' should be put together from the very start of the project and reflect the issue(s) at hand, the date issue(s) identified, the problem, recommendations, and the impact on the project if the issue is not resolved.

Where many divisions are implementing the system and integration is required across the divisions, the problems can become overwhelming. These issues can be many and varied, but key point here is that the issues log is kept in front of and discussed daily, if necessary, with the CEO or delegated senior executive that has the authority to make or have decisions made to resolve the issues. 'See what you can do or see what happens' is not a resolution and may well come back to cause considerable problems further down the line and even system collapse when the system is turned on to live running.

Failure to resolve issues or putting them in the too hard basket can have a demoralising effect on the entire project team that is tasked with the job of implementing the system and provide any third-party contractors with the perfect excuse for non-performing or failed systems!

21. Scope Change

ERP projects begin life with a set of objectives, some specific, some vague. To support these objectives, a definition of the project is made and a specification for what they expect the ERP software to do is defined. The software is selected after demonstrations and negotiations on support, pricing, etc., and at some point, the ERP project commences.

A point worth noting is that a major contributor to ERP project failures is the selection of an ERP software package that is simply the wrong package for the intended use. This may be lack of functionality, lack of integration, too convoluted to use effectively, or simply unsuitable for the target business. You may ask why this should happen if we specified and demonstrated the ERP software and it appeared to work. The reality in most organisations is that the demonstrations may well have been done piecemeal, that is, looking at different parts of the system in isolation and not as a fully integrated system or you have been cleverly misled by the software house that deliberately did not show you elements of the system that may have jeopardised their prospects of a sale.

There are also issues with the lack of internal discipline that allows employees to try to get the system to duplicate the functionality of their existing systems without realising the capability of the new ERP software and try to change the new system to do what they know and are comfortable with.

Scope creep is a gold mine for ERP vendors and their implementation teams or partners, and it involves the buying organisation asking for additional functionality or modifications to the software and in many cases the purchase of additional modules in order to achieve the required functionality. ERP vendor's implementation teams are given kudos for expanding the revenue opportunities within client businesses, and it is commonly called the ERP gravy train. If an organisation asks for any changes to software or scope, and they are subtly encouraged to do so by the software vendor's consultants, then watch the money-go-round spin and the project time-line and budget blowout.

The best way to avoid selecting the wrong software is for an organisation to *develop a model* together with any specifics in each function identified and then test the ERP software by simulating it through the model together with the specifics in each function required!

22. Software Changes

Changes to the new ERP software post-purchase during the implementation stage of the software, whilst generally willingly accepted by the ERP software vendor, will be costly and disruptive to the overall implementation effort.

All requests for modifications or changes to the software should be logged and a business case put forward for evaluation and approval by the CEO before any changes are approved. This will have the effect of eliminating frivolous changes and reduce the need for costly software modification and project time delays whilst ensuring that any changes are approved are really required.

It is also worthwhile putting requested changes into two categories: (1) essential to implementation effort and (2) to be reassessed post-live running of the system. It is not unusual for modifications classified as category 2 to be found unnecessary after the system is implemented as there are tools existing within the system to achieve the desired functionality.

ERP software vendors are very happy to modify and change software and typically expect companies to ask them to undertake changes during the implementation stage of the project. The cost of these changes adds to the revenue stream of the software house, can be very costly, delays the project, and contractually gives the ERP vendors team reasons to blame the client for delays. In addition to the costs and delays, ERP vendors will, usually, not agree to maintenance on the software changes, and any ongoing software upgrades become costly and problematic, as the software changes have to have special modifications done to fit the previous non-standard modifications the organisation requested.

Changes to software should be kept to a minimum if the software is simulated through a model, and any changes required should be agreed to before the software is purchased and should include guarantees on additional ongoing maintenance at no cost to the client.

Another issue worth noting is that even if an organisation pays for software modifications, the ERP vendor typically retains ownership of the modifications and is free to sell the modifications to your competitors, even though you have paid for the development of the changes.

These issues need to be agreed on and resolved at the negotiation stage and clearly stated in the contract who owns the intellectual property of the software enhancements, paid for by the client.

Ensuring a close fit via the model simulation during the ERP software evaluation stage and then controlling any changes are the best way for an organisation to protect themselves from escalating costs of software changes, project impact, ongoing maintenance, and upgrade issues.

23. Management Action on Issues

Ongoing management involvement in the ERP project is typically via some form of steering oversight committee made up of the very senior level of the organisation and chaired by the CEO or most senior person onsite. The project manager should have access to the CEO on a daily basis to discuss any issues that affect the progress of the project, and the steering committee should review the project on a weekly basis for any major issues that require resolution.

Part of the reason senior management fail to get involved at a significant level in the project is a lack of understanding of what the organisation is trying to achieve and the tendency to relegate the project to the status of a computer system. When difficult decisions need to be made by senior management on issues encountered during the ERP project cycle, the tendency is to put the problem in the too hard basket or put off making a decision in the hope the issue will resolve itself. This causes disruption to the project and a loss of confidence that management is prepared to get involved to resolve issues.

Management involvement has been cited as a major impediment to the successful implementation of an ERP project! An education program involving the organisation, including senior executives and management, will overcome many downstream hurdles through an understanding of the technology and changes that are required to achieve a successful ERP outcome.

An issues log, raised by the project manager, must be dealt with and a resolution found, whatever that solution is!

An ERP project will cross every organisational boundary, and the only person in the business who can make decisions across these boundaries is the CEO as no other person has been given a wide enough mandate to make changes and resolve the issues arising from those changes. Resolving these issues may be problematic as typically it requires change that may affect other executives and departments.

Failure of management to get involved will be used by software houses and their implementation partners as a prime reason for the systems' underperforming or failure when switched on and a justification for delays and cost blowouts on the project. In addition, the morale of the project manager and their team and the entire organisation can be affected as the ERP project bogs down through lack of progress and direction.

24. Go-Live Readiness Reviews

The live running of an ERP system is the ultimate test of all the work that has taken place with education, configuration, training, data clean-up, data migration, process re-engineering, procedures, and change management.

The testing of each function for processing and integration for sign-off prior to live running should identify issues missed during the activities of the implementation stage of the project.

Short cuts taken during the implementation process and the live running readiness reviews can cause major issues when the system is cutover to live running.

These reviews, also commonly referred to as conference room pilots, are conducted simulating full integration, in detail, on how the system will perform when the system is switched to live running. This is the opportunity to identify any shortcomings and rectify them without causing major disruption to the organisation or in some cases processing collapse and chaos when the system is turned on.

There are potential traps in thinking that a successful live running readiness review will guarantee a good ERP outcome when you go live! What an organisation has done with a successful live running readiness review is simulate and test the system through the processes of the business, through a number of test scripts, to ensure the system works end beginning to end to ensure it integrates.

The issue here is one of the data accuracy and integrity outside the data sets used in the readiness review simulation. In the review, simulation sets of data were used to test the system for processing and integration.

These data sets may have been taken from data in the system or data put together just for the review and may not reflect the accuracy of the underlying data that resides in the data files migrated into the system from legacy systems.

This can be a disaster waiting to happen when the system is turned on!

Assumptions on data accuracy may be incorrect, and the data accuracy of the migrated data must be subject to rigorous checks before signing off for live running.

The review is to test the system to ensure it processes the data through the ERP system using data sets that may or may not be correct. There is no guarantee the data was correct in the first place as that is not the focus of the readiness review test.

The ERP system will process correct or incorrect data and has no way of knowing if the base data is correct. For example, a bill of material may have been used that has the wrong components, be over—or under-structured, be missing assemblies or sub-assemblies, or has a number of other issues that will impact the effective processing of any MRP output. The costing information may be incorrect, but the system will process both the bill of material and calculate the cost information correctly from a software perspective based on the incorrect data and look as though the system is working. It is! But it may not reflect the issues that will be encountered in the real world due to the problems with the real live data.

The system may well work from a software simulation and integration perspective, but if information in the data required for live running is incorrect, the outputs will not support the activities of the supply chain and cause operational chaos and potential disaster when the system goes to live running status.

This goes back to the project activities where the data clean-up and migration may have been less than diligent or short cuts taken to make up for project delays with the view that the data can be cleaned up post-live running. The go-live readiness review may well create a false confidence which comes crashing down when the system goes live and the outputs based on poor data cannot be used by the organisation.

Companies have reported significant negative impacts on operations, customer service, lost sales, production, costs, and even bankruptcy as a result of systems passing readiness reviews and go-live decisions signed off but had data issues that were not resolved during the run-up to implementation.

Care must be taken to ensure all the clean-up and data migration work on data are completed accurately and signed off before the system goes live or face potentially disastrous consequences when the system goes live.

25. Live Running Cutover

At some point, the ERP system will be turned on in the live environment. There are typically three approaches that companies make on live cutover: *big bang* (turn it all on at once), *parallel running* (run it alongside existing systems until you have confidence in the new ERP), and *phased turn-on* (turn parts of it on and add more when confident).

Whichever method chosen, there are arguments for and against each. The results will depend on the diligence of work done in the run-up to live running. Tasks missed or poorly carried out in the project task stage will inevitably cause problems once live running is cutover, and if enough tasks have been poorly carried out, then operational problems will occur.

The method of live running will depend on the difficulty created by each method. Discussions within the company and the implications of each method should be examined before the decision on which method of live running is agreed on.

The issues identified should be documented and risk strategies put into place to cover any problems that may arise!

From a senior executive's perspective, close scrutiny of the project's progress and knowledge of all the issues that have a potential to cause post-live running problems need to be *assessed for risk* to the business before the system is switched on.

This is a critical period for the company, and a 'wing-it' approach must not be accepted. It is better to delay live running to ensure every element of the system is correct, including procedures, what reports are required, system access rights, who is looking after what, and that all the users have been trained so that work can proceed smoothly when the system is turned on. This is a critical time for the entire project, and management's involvement at each stage of the project must ensure that decisions made, which may have an impact on the ultimate outcome, are understood and responsibility accepted for those decisions.

26. Post-ERP Cutover

Going live with the ERP system is a critical time for all organisations. The work carried out on all the implementation activities will come to fruition when the organisation attempts to use the system in the real live environment. Any short cuts taken during the implementation stage without a full analysis of the impact on live running, inadequate training, data clean-up and migration, lack of access for users, security, data issues, poorly defined procedures, confusion on output reports and their interpretation, lack of system operational measurements, etc., will all impact the organisation's ability to manage the business and can seriously disrupt the business.

Results range from minor issues to major processing problems, preventing shipping of goods, scheduling, processing orders, inventory corruption, financial errors and loss of control resulting in significant losses, and in some cases bankruptcy.

The go-live readiness review carried out prior to committing to the live running, if done effectively, should have filtered out the potential problems and if necessary delayed the live running cutover.

The project doesn't end with the system being cutover to live running. A small team should be on standby to deal with any issues that may occur in the first weeks of the systems operations. Any issues should be logged and management made aware of the problems via a review of the live running issues log, which must be followed up on to ensure all issues are resolved.

Typically, the cutover to the ERP live running will cause some initial disruption until the users get their heads around the smooth operation of the system.

Management should conduct weekly reviews of the system's operation to ensure there are no major impacts on the business. A measurement system should also be in place to determine if the company is meeting the outcomes cost-justified for in the upfront acquisition rationale.

Final Comments

ERP is a technology that has promised great benefits to organisations! The theory of the technology is sound, and some organisations have gained great benefits from the technology and use it as a strategic tool to manage and grow their organisations.

The greater majority of organisations struggle with the implementation of the technology and have outcomes ranging from mediocre to disaster.

There is no such thing as an easy implementation of an ERP project. Easy implementation is simply sales hype pushed by ERP software vendors.

With the ERP industry being worth $50 billion, a whole culture has evolved to take a slice of the business. The practices of the ERP sellers, their implementation teams, and the raft of so-called expert consultants have managed to create an air of necessity for their services, to their own advantage. The recipients and victims in all of this are the buying organisations which sign on to the sales hype only to find that ERP vendors and their implantation teams take little or no responsibility or accountability for delivering outcomes and simply walk away leaving the legacy of whatever outcome is achieved.

Ultimately, the difference between successful ERP project outcomes and failure is the approach senior executives take towards the project and their willingness to get involved at each stage to ensure success.

Organisational ownership to ensure the technology and the issues that need to be addressed to achieve a successful outcome can be obtained through effective education prior to launching into an ERP project. This education will ultimately be the difference between success and failure for any organisation.

Senior executives and management are not expected to become technical experts on ERP. They are however required to understand the issues the company will face when embarking on an ERP acquisition and implementation.

This knowledge is generally lacking across most ERP projects, and unless there is a change in the way ERP projects are viewed and managed, the efforts to modernise and utilise the ERP technology to the benefit of our industries and organisations will continue to elude us.

The 26 steps we have outlined here are a logical progression of a project and the issues that will be encountered.

The expectation that the ERP industry will become better and more accountable is not supported by 30 years of poor outcomes.

With the huge sums of money and potential impacts on the revenue of these ERP sellers and their implementation services, it is unlikely they will change unless the customers become more knowledgeable and force the changes necessary.

Our industries are facing ever increasing competition globally and the only way we can remain competitive and satisfy demands of customers on a cost effective basis is to adopt and utilise technology quickly and successfully.

ERP has evolved from the 1960s to the systems we have today. It is inconceivable that the level of failures being experienced in ERP projects would be acceptable in any other undertaking in the history of human endeavour. We are intelligent beings that are able to innovate, identify problems and overcome them to achieve our goals. Why then after thirty years have we not achieved the same result with ERP systems? The experience is there, we know what constitutes success and failure. We simply have to change the approach and understand that the solution is not in paying large sums of money for failure but in investing in education, training and internal ownership as the way to successful ERP outcomes. *This book was never intended as a 'how to' implement an ERP system but is aimed at giving senior executives and management an understanding of the issues they will encounter when embarking on an ERP acquisition and implementation project.*

* * *

Index

A

accountability   30, 47, 78
acquisition rationale   34-6, 40

B

bankruptcy   16, 25, 27, 75-6
blame game   19, 22, 24
budget   9, 21, 34, 38, 57, 59, 61, 66
business   9, 12-13, 21, 26, 35-40, 43, 45, 49-51, 55, 58, 60, 65, 72-3, 76-8
business processes   63-4
buyer   22-3, 29, 60

C

cash cow   59, 61
client   10, 22-3, 27-9, 53, 58-62, 70-1
 new   58
 potential   60
 prospective   59
 unhappy   58
client businesses   69
client companies   28, 31
client organisations   62
complaints   20-1, 45-6, 51, 67

computer system   11-12, 24, 55, 71
computers   9, 11, 26
consultants   28-30, 56, 58-9, 61-3
 independent   46, 48
contracts   23, 28, 31, 47, 59-60, 71
costs   9, 13-15, 20-1, 27-9, 37-41, 45, 50, 53, 59-60, 66, 70, 75, 79
CRM (customer relationship management)   13

D

data accuracy   51, 73-4
data cleanup   21, 62, 66, 72, 74, 76
data conversion   31, 48, 66-7
data sets   73-4
deliverables   46-7
demonstrations   29, 68-9
Devil's Triangle   21, 26, 53

E

education   12-13, 21, 31, 37, 50-1, 57, 72, 78-9
ERP (ENTERPRISE RESOURCE

PLANNING) 9-11, 13-15, 17, 21, 23, 26-7, 29, 31, 33-9, 41, 43, 45, 47, 49-51, 53-5, 57, 59-61, 63-5, 67, 69, 71, 73, 75, 77-9
ERP acquisition 35, 50, 78-9
ERP-buying company 48
ERP clients 57
ERP failure 10, 20, 23, 26, 38, 56
  major 18
ERP implementation process 50, 58
ERP implementation projects 51
ERP implementations 19, 27, 46, 54, 59, 64, 66
ERP industry 9-10, 25-7, 30-1, 57, 62-3, 79
ERP information system 18
ERP project failures 26, 68
ERP project manager 64
ERP project plan 48, 54
ERP projects 9-10, 14-15, 17-20, 22-3, 25-6, 30, 32-8, 40, 42, 46, 49, 53-5, 57, 60, 62, 64-5, 67-8, 71-2, 77-9
ERP software 17, 40, 46, 67-9
ERP software investment 65
ERP software package 66, 68
ERP software providers 51-2
ERP software supplier 23, 25, 52, 58
ERP software vendors 20-1, 23, 25-6, 41, 47-8, 53, 58, 62, 70, 77
ERP systems 15-16, 26-7, 34-7, 40, 50, 56, 62-5, 72, 74-6, 79
ERP systems implementation 24
ERP technology 15, 25, 78
ERP type systems 63
ERP vendors 9, 22-3, 26-7, 30, 33-4, 37, 45, 47-8, 54, 59-61, 66, 69-71, 78
executive involvement 56-7, 67
expertise 20, 22-3, 26, 28, 36-7, 45, 54, 63

F

failure rates 10, 27, 60
functionality 28-9, 32, 40, 42, 44, 61, 68-70
functions 9, 11, 38, 43, 52, 63, 69, 73

G

go-live readiness reviews 74, 77
goodwill 46

H

HR (human resources) 13

I

implementation 9, 12-14, 17, 22, 24, 27, 29, 37, 45, 49, 53-4, 58-9, 61-2, 67, 72, 75, 77-8
  failed 26, 40
  successful 14, 27, 61, 72
implementation activities 49, 76
implementation consultants 27
implementation effort 14, 49, 70
implementation partners 19, 21-3, 25-7, 30, 37, 54, 58, 62, 72
implementation plans 50, 53
implementation process 48, 54, 58-61, 66, 73
implementation project 54, 79
implementation services 13-

14, 45, 59, 79
implementation stage 28, 40, 42, 44, 46, 59, 66, 69-70, 73, 76
implementation teams 45-7, 53, 69, 78
  vendor's 53-4
industries 10, 13, 15, 22-3, 25-6, 40, 42, 45, 78-9
information 15, 18, 36-7, 40, 46, 74
  real-time 9, 15
integration 44-5, 68, 73
investment 14, 16, 59

J

jargon 42, 59-60

L

legacy systems 39, 65-6, 73
litigation 20, 23, 30
live running cutover 77

M

management 15, 17-18, 20, 26, 28, 33-4, 36, 52, 55-7, 71-2, 77-9
milestones 23, 47, 49, 63
modifications 29, 42, 53, 59, 69-71
money 10, 14, 22, 25-30, 41, 50, 56, 61-2
MRP (materials requirements planning) 9, 11, 16, 50, 56

O

organisations
  boutique 62
  manufacturing 36
ownership, internal client project 27

P

partners 9, 14, 19-20, 22, 25-7, 30-1, 45, 47-8, 53, 58, 62, 69
products 11-12, 26, 41-2, 44, 47, 65
project 13-16, 18-20, 24, 26-9, 31-8, 40, 42, 46, 48-50, 52-62, 66-8, 70-3, 76-8
  business transformation 55
project activities 47, 54, 56, 63, 74
project management 10, 48, 55, 62
project management team 34
project manager 10, 49, 55-7, 67, 71-2
  in-house 48
  internal 24
  third-party 48
project plan 14, 23, 29, 48-9, 54-7
project teams 10, 57, 67
project time delays 29, 70

R

resources 18, 46
  organisation's 18
responsibility 10, 21, 23-4, 26, 30, 34, 47-8, 53-4, 56, 58, 76, 78
revenue 25, 29, 31, 37, 53, 63, 79
RFP (request for proposal) 33, 40-1

S

sales hype 20, 42, 47, 77-8
SAP 16-17, 30
selection 44, 46, 68
services 14, 31, 39, 45-9, 53, 60, 62, 78

simulation   41
software   11, 13, 17, 19, 26-8, 32, 34-6, 39-44, 46, 48-9, 54-5, 57-60, 62, 66, 68-70
  best practice   42
  data conversion   66
software acquisition   13, 46
software budget   28
software changes   34, 70-1
software companies   13-14, 19-20, 29, 45
software component   42
software configuration   41, 64
software consultants   34
software cost   45, 61
software enhancements   71
software functionality   20, 28, 46, 48
software functions   51
software house translation   60
software houses   19, 21-2, 29-32, 46, 49, 58-9, 62, 69-70, 72
software implementation process   65
software inadequacies   40
software industry   25
software modifications   71
software modules   36, 61
software package   61
software providers   13, 30, 40
software sales process   48
software simulation   74
software suitability   42
software suppliers   27, 57
software technology   10
software tools   35
software training   51
software upgrades   70
software vendors   9, 21, 27, 31, 46, 59, 62
success   10, 12, 14-15, 19, 21, 32-3, 36-7, 46, 54, 57, 60, 78

system   11-13, 18, 21, 23, 26, 35, 40-1, 45, 50, 52, 55-6, 58, 62, 65-70, 72-7, 79
system capability   23
system users   57

T

technology   10-14, 18-19, 22, 26-8, 31, 37, 39, 42, 44, 48-50, 72, 77-9
  add-ons   13
  capability   18
  component   48-9
  implementation   55
  project   15, 35-6, 67
  software component   48
timescales   21, 23
training   13, 20-1, 37, 48-9, 51-2, 62, 72, 79
training ground   28, 58

U

users   11, 21, 40, 45, 51-2, 67, 76-7
  super   52

V

vendor   20, 22-3, 33-4, 41-2, 61
vendor implementation services   46
vendors implementation team   53
vision   53

W

Waste Management   17
Whirlpool   16-17

www.ingramcontent.com/pod-product-compliance
Lightning Source LLC
Chambersburg PA
CBHW022126170526
45157CB00004B/1769